SAE EDGE™
RESEARCH REPORT

Unsettled Topics Concerning Adopting Blockchain Technology in Aerospace

Rhonda D. Walthall

Collins Aerospace

EDGE DEVELOPMENT TEAM

Tim Abbott, *Moog Aircraft Group*

Brad balance, *Airlines for America*

Stylianos Basagiannis, Ph.D., *Raytheon Technologies Research Center*

Arnaud Brolly, *SITA*

Dragos Budeanu, *International Air Transport Association*

Srikanth Challa, *Infosys*

Paul Conn, *Airlines for America*

Aharon David, *AFUZION-InfoSec*

Chris Fabre, *Sky Republic*

Stefan Fölser, *eWINGZ*

Leon Gommans, Ph.D., *KLM*

Fred Jones, *Raytheon Technologies Corporation*

Jason Jones, *Moog Inc.*

Ken Jones, *Airlines for America*

Scott Kordella, Sc.D., *MITRE Corporation*

Kevin Kuczynski, *Pennsylvania State University Applied Research Lab*

Ravi Kumar G.V.V., Ph.D., *Infosys*

Chris Markou, Ph.D., *International Air Transport Association*

Chuck Marx (Ret.), *Digital Aerospace Consulting (PwC)*

Sean Melia, *SITA*

Ravi Rajamani, Ph.D., *drR2 consulting*

Harvey Reed, *MITRE Corporation*

Rusty Rentsch, *Aerospace Industries Association*

Col. James Regenor (jar), *VeriTX*

Mark Roboff, *DXC Technology*

Adam Siena, *Block Aero*

Todd Siena, *Block Aero*

Aaron Spak, *Pennsylvania State University Applied Research Lab*

W. Ben Towne, Ph.D., *SAE International*

Martin Whitfield, *SAP Corporation*

SAE INTERNATIONAL

Warrendale, Pennsylvania, USA

About the Publisher

SAE International® is a global association of more than 128,000 engineers and related technical experts in the aerospace, automotive, and commercial-vehicle industries. Our core competencies are life-long learning and voluntary consensus standards development. Visit sae.org.

SAE EDGE™ Research Report Disclaimer

SAE EDGE™ Research Reports focus on topics that are dynamic, in which knowledge is incomplete, and which have yet to be standardized. They represent the collective wisdom of a group of experts and serve as a practical guide to the reader in understanding unsettled subject matter. They are not meant to provide a recommended practice or protocol. The experts have assembled as a community of practitioners to contribute and collectivize their thoughts and points of view; these are not the positions of the institutions or businesses with which they are affiliated, nor is one contributor's perspective advanced over other contributors. SAE EDGE™ Research Reports are the property of SAE International and SAE alone is responsible for their content.

About This Publication

SAE EDGE™ Research Reports provide state-of-the-art and state-of-industry examinations of the most significant topics in mobility engineering. SAE EDGE™ contributors are experts from research, academia, and industry, who have come together to explore and define the most critical advancements, challenges, and future direction in areas such as vehicle automation, unmanned aircraft, cybersecurity, advanced propulsion, advanced manufacturing, Internet of Things, and connectivity.

Related Resources

SAE Mobilus® Cybersecurity Knowledge Hub
https://saemobilus.sae.org/cybersecurity/

SAE Mobilus® Automated & Connected Knowledge Hub
https://saemobilus.sae.org/automated-connected/

SAE Team

Frank Menchaca, Chief Growth Officer
Michael Thompson, Director of Standards, Information and Research Publications
Monica Nogueira, Director of Content Acquisition
Beth Ellen Dibeler, Product Manager
William Kucinski, Managing Technical Editor

EPR2020021
ISSN 2640-3536
e-ISSN 2640-3544
ISBN 978-1-4686-0250-0

To purchase bulk quantities, please contact: SAE Customer Service

E-mail: CustomerService@sae.org
Phone: 877-606-7323 (*inside USA and Canada*)
         +1-724-776-4970 (*outside USA*)
Fax: +1-724-776-0790

https://www.sae.org/publications/edge-research-reports

About the Editor

Rhonda D. Walthall is a Technical Fellow at Collins Aerospace in Charlotte, NC, a division of Raytheon Technologies Corporation. In her role, she focuses on Design for Prognostics and Health Management (PHM) and Digital Thread initiatives. She is an industry-recognized leader in the development of standards and best practices for Integrated Aircraft Health Management (IAHM).

Walthall earned her Bachelor of Science degree in Aeronautical and Astronautical Engineering from Purdue University and her Master's degree in Business Administration from Pepperdine University. She began her career as a Flight Test Engineer for the McDonnell Douglas Aircraft Company before working for Northwest Airlines as a Powerplant Engineer. In 2003, she joined Hamilton Sundstrand and has held positions of increasing responsibility, leading to her role today as a Technical Fellow.

Walthall is a Fellow and Vice President of the PHM Society. She is a member of the SAE International Board of Directors and has been actively engaged in SAE standards development and leadership since 2004. She was recognized by SAE as a Top Contributor of the Year (2019) and received the Rodica Baranescu Award for Technical Excellence & Leadership (2018) and the James M. Crawford Technical Standards Award for Outstanding Achievement (2016). She holds three patents, has contributed significantly to numerous SAE documents, and authored chapters in SAE and IEEE publications.

In 2020, Rhonda received the Outstanding Aerospace Engineer Award from Purdue University, where she is a member of the Industrial Advisory Council to the School of Aeronautics and Astronautics.

contents

Atosan/Shutterstock.com

Unsettled Topics Concerning Adopting Blockchain Technology in Aerospace

Abstract

Aerospace is an industry where competition is high and the need to ensure safety and security while managing costs is foremost. Stakeholders, who gain the most by working together, do not necessarily trust each other. Changing backbone technologies that drive enterprise systems and secure historical records does not happen quickly (if at all). At best, businesses adapt incrementally, building customized applications on top of legacy systems. The complexity of these legacy systems leads to duplication of efforts and data storage, making them very inefficient. Technology that augments, rather than replaces, is needed to transform these complex systems into efficient, digital processes.

Blockchain technology offers collaborative opportunities for solving some of the data problems that have long challenged the aerospace industry. The industry has been slow to adopt the technology even though experts agree that it has real potential to revolutionize the global supply chain—including maintenance, repair, and overhaul (MRO)—driving tremendous cost, excess inventory, and inefficiencies out of the system. This report discusses how the adoption of blockchain technology could have a significant impact on the aerospace industry and addresses some of the unsettled concerns surrounding the implementation of the technology.

NOTE: SAE EDGE™ Research Reports are intended to identify and illuminate key issues in emerging, but still unsettled, technologies of interest to the mobility industry. The goal of SAE EDGE™ Research Reports is to stimulate discussion and work in the hope of promoting and speeding resolution of identified issues. SAE EDGE™ Research Reports are not intended to resolve the challenges they identify or close any topic to further scrutiny.

RHONDA D. WALTHALL
Collins Aerospace

Edge Development Team

Tim Abbott, *Moog Aircraft Group*
Brad balance, *Airlines for America*
Stylianos Basagiannis, Ph.D., *Raytheon Technologies Research Center*
Arnaud Brolly, *SITA*
Dragos Budeanu, *International Air Transport Association*
Srikanth Challa, *Infosys*
Paul Conn, *Airlines for America*
Aharon David, *AFUZION-InfoSec*
Chris Fabre, *Sky Republic*
Stefan Fölser, *eWINGZ*
Leon Gommans, Ph.D., *KLM*
Fred Jones, *Raytheon Technologies Corporation*
Jason Jones, *Moog Inc.*
Ken Jones, *Airlines for America*
Scott Kordella, Sc.D., *MITRE Corporation*
Kevin Kuczynski, *Pennsylvania State University Applied Research Lab*
Ravi Kumar G.V.V., Ph.D., *Infosys*
Chris Markou, Ph.D., *International Air Transport Association*
Chuck Marx (Ret.), *Digital Aerospace Consulting (PwC)*
Sean Melia, *SITA*
Ravi Rajamani, Ph.D., *drR2 consulting*
Harvey Reed, *MITRE Corporation*
Rusty Rentsch, *Aerospace Industries Association*
Col. James Regenor (jar), *VeriTX*
Mark Roboff, *DXC Technology*
Adam Siena, *Block Aero*
Todd Siena, *Block Aero*
Aaron Spak, *Pennsylvania State University Applied Research Lab*
W. Ben Towne, Ph.D., *SAE International*
Martin Whitfield, *SAP Corporation*

ISSN 2640-3536

3

Introduction

State of the Industry

Commercial aerospace has enjoyed nearly a decade of profitability, prompting most businesses to invest in technologies that would improve efficiency, including upgrading their enterprise systems and infrastructures and deploying data analytics to gain business insights. This period of investment has nudged the industry forward toward digital transformation. Abruptly, the COVID-19 pandemic stopped the industry in its path and drove airlines to accelerate the retirement of older aircraft, return leased aircraft, and cancel orders for new ones. As aircraft were parked, the need for maintenance and spare parts declined dramatically. The ripple effect throughout the entire supply chain has been, and will continue to be for many years, devastating.

While the length of the recovery period is uncertain, the timing is right to implement transformative technologies that are at the heart of Industry 4.0. Industry experts remain optimistic about the adoption and implementation of blockchain technology and some say it is not a question of "not if, but when" the technology will be widely adopted. Blockchain technology has the potential to remove inefficiencies throughout the aerospace ecosystem, reduce the cost of operations and maintenance, secure communications transmissions, and support initiatives that will have a long-term positive impact on the environment. To achieve this potential, industry leaders must first agree that the time for change is now. They must understand the technology and align on use cases. They must agree to governance structures and commit to sharing the cost of implementation. Ensuring the adoption of blockchain technology will require no less than a wide-scale industry collaboration.

How Does Blockchain Technology Work? A blockchain consists of blocks of time-stamped information (data, transactions, records, etc.) that are linked together by the current hash (the output of an algorithm generated from the information within the current block and the hash of the previous block of information). When a transaction is completed, that transaction must be verified by all stakeholders (nodes) as being valid and complete. Once a consensus is reached by the stakeholders that the transaction is valid and complete, the block containing the transaction is appended to the blockchain. Each node then maintains an identical copy of the entire blockchain, which makes it difficult for a malicious actor to alter any information in the blockchain. In order to do so, the bad actor would have to alter a transaction in a given block, recalculate the hash of all the subsequent blocks, and then achieve consensus across the distributed nodes in that version of the blockchain was the most authentic version, including the altered block.

Blockchain technology started with cryptocurrency and finance. However, it has evolved over the past decade and now includes distributed ledger technology (DLT), which is simply a database (ledger) that exists across multiple independent locations and participants (nodes) [1]. The information stored in the ledgers is synchronized so that each copy of the ledger contains the same set of information (transaction records). Blockchain technology builds upon DLT by organizing the transaction records into blocks and then linking them together cryptographically so that the records are resistant to tampering. Many industries, including agriculture, automotive, consumer products, medical, pharmaceutical, and luxury retail, use blockchain technology today.

How Has Blockchain Technology Been Explored in Aerospace? Since 2015, the aerospace industry has been stuck in "Proof of Concept Purgatory" [2]. Companies of all sizes involved in manufacturing, operating, repairing, leasing, selling, insuring, supporting, and certifying aircraft have started to explore use cases for this disruptive technology. Numerous white papers have been published espousing the benefits of the technology when used for specific purposes with narrow solutions. Ideation workshops have been held around the world to develop and advance blockchain concepts. Software companies, such as IBM, Accenture, and SAP Corporation, have developed applications that could be used with blockchain technology to fill the unique needs of the industry. Startups have emerged with software-as-a-service (SaaS) solutions to help businesses adopt and deploy a blockchain network. Consortia and alliances have been formed between major entities.

A few successful blockchain trials have pushed commercial adoption forward. For example, in 2018, the first all-digital transaction was demonstrated by ST Aerospace and Moog. ST Aerospace purchased a digital design of a part from Moog, and then used additive manufacturing (AM) technology to create the part in its facility in Singapore. The financial settlement between the two companies was completed instantaneously using a smart contract on the Microsoft Azure Blockchain [3]. On April 9, 2019, Air New Zealand, Moog, ST Aerospace, and Microsoft demonstrated that a commercial seat part could be produced using AM (otherwise known as "3D printing") while the aircraft—a Boeing 777—was enroute from Auckland to Los Angeles. As soon as the aircraft landed, the new part was installed without causing a flight delay [4]. More recently, several aerospace companies, such as Honeywell [5], Boeing [6], Airbus [7], and Moog, announced that they were launching commercial services utilizing blockchain technology.

The United States military has taken a more aggressive approach to exploring blockchain technology. The 2018 National Defense Authorization Act included a provision that ordered the Department of Defense (DoD) to conduct a comprehensive study of blockchain, with an emphasis on cybersecurity [8]. In March 2019, the DoD announced that five blockchain demonstrations had been conducted at five different depots using the Moog VeriPart® solution and Guardtime Federal Core Blockchain. The DoD announced that they had successfully demonstrated the ability to use the technology to complete remote AM, mitigate counterfeit parts, execute smart contracts, ensure parts provenance, validate process integrity, audit ledgers, and perform build analyses against part designs [9]. Further, the DoD Information Resource Management Strategic Plan FY19-23 included two blockchain projects: one focusing on securing sensitive communications and one focusing on cybersecurity of databases [10].

Even though the industry has been exploring the technology for several years, the potential benefits have not been realized because of the low adoption rate. This SAE EDGE™ Research Report seeks to explore the unsettled issues concerning the adoption of blockchain technology in the aerospace industry, specifically identifying which use cases would solve the biggest challenges, overcoming hurdles related to perception and standardization, and determining the best implementation and governance strategy to achieve the desired objectives.

Solving the Greatest Challenges

In October 2018, a survey conducted by the Boston Consulting Group (BCG) and Aerospace Industries Association (AIA) found that 20% of aerospace and defense supply chain respondents indicated that their companies were assessing blockchain technology [11]. By January 2020, a survey conducted by Accenture indicated that 61% of aerospace and defense companies were implementing or piloting a DLT-like blockchain [12]. What inspired this dramatic increase in interest? Hype? Fear of missing out? Record profits? Insights gained through Proof of Concept projects? The answer is likely "Yes" to all of these.

If the Accenture survey is indicative of reality, then most aerospace and defense leaders believe that there is true value to be gained with blockchain technology. But is this truly reality?

Only a handful of aerospace companies have reached the point where they perceive enough value to move forward with a blockchain platform implementation. Most have chosen to form partnerships, alliances, or consortia in order to achieve a specific purpose. Some of the notable platforms available today are: Aeron, ARINC, Block Aero Aviation Blockchain Platform, Ethereum, Honeywell's GoDirect Trade and iTRACE, Hyperledger, Lufthansa's Blockchain for Aviation or "BC4A," MRO Blockchain Alliance (lead by SITA), Moog's VeriPart®, SITA's FlightChain, Zamna, and the SAE ITC® TS200™ Database used by the Aerospace Standards Part Qualification Program.

When the contributors of this SAE EDGE Research Report were asked what the greatest benefits were in using blockchain technology, their top two responses were: (1) trust and (2) visibility into transactions/transparency (Figure 1). How do these two benefits bring value? Does a lack of trust and transparency

FIGURE 1. Benefits of using blockchain applications in aerospace.

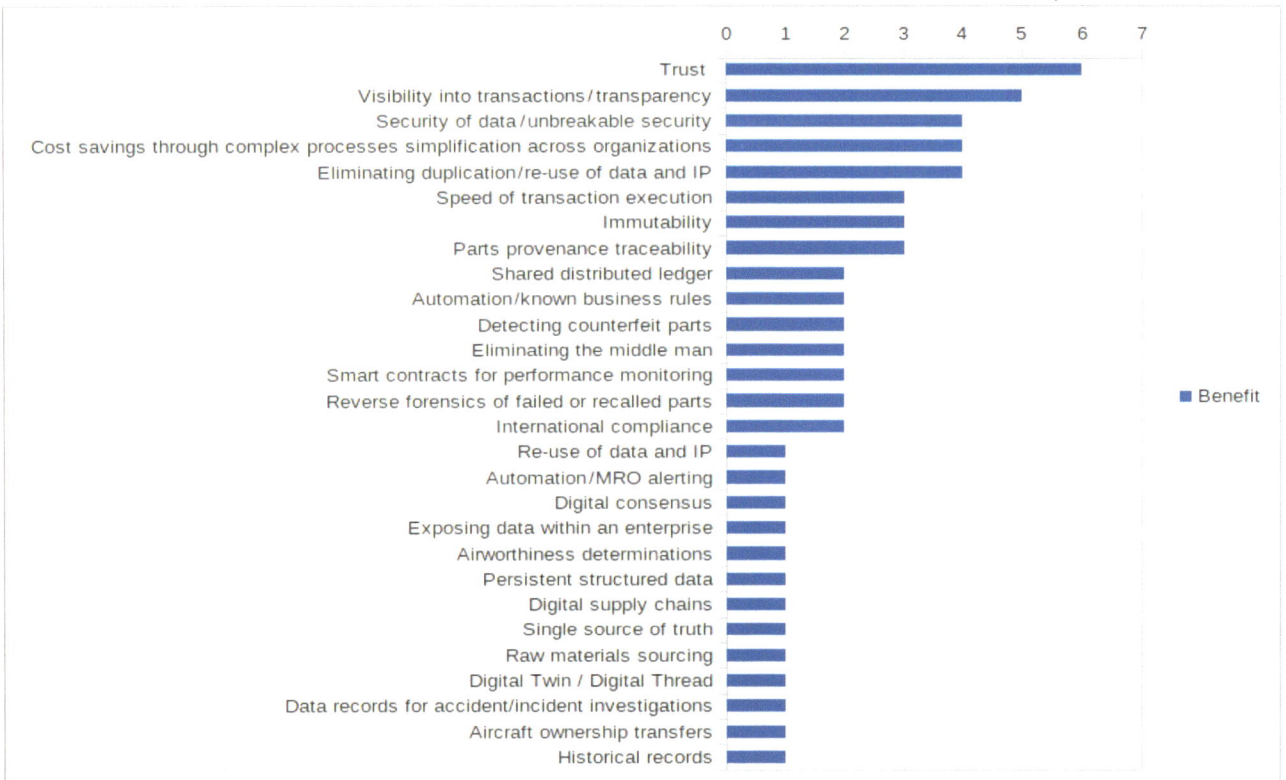

© SAE International.

in today's business transactions drive a need for a blockchain solution? Or, does confidence in the integrity of the information in the blockchain—eliminating the need to verify and duplicate the information—drive the true value? More importantly, which use cases would derive the most value by providing a solution to the biggest challenges facing the industry?

The pace at which the industry is moving from paper-based transactions to electronic transactions is continuing at an increasing rate. Some of the use cases described in this report are related to this transformation, which means that the unique advantages of blockchain technology are not what is driving the change. This "digital transformation" will happen regardless of the underlying database technology used to store the resulting electronic data. Businesses are already working with centralized databases that have adequate data security, redundancy, access controls, and so forth for storing and tracking data. However, in certain cases, blockchain technology could present a better alternative, especially if there are multiple stakeholders involved, or multiple servers are needed to store transaction records, or the process itself becomes too complicated to develop contracts for data storage and usage by various stakeholders. An example of this would be in maintenance, repair, and overhaul (MRO) ecosystem.

In 2018, International Air Transport Association (IATA) published a white paper that highlighted the results of a five-year study of the "Future of the Airline Industry 2035." The white paper described the potential opportunities that blockchain technology could offer the airline industry [13]. The contributors to this SAE EDGE Research Report agreed with the widely recognized use cases described within the white paper as well as in other published research, but identified several more novel use cases that offered unconventional perspectives on how blockchain technology could be used to solve existing as well as expected problems in the future (Table 1).

Supply Chain Challenges

Many of the greatest challenges facing the aerospace industry revolve around inefficiencies in the supply chain that result from lack of visibility, duplication of efforts, human errors, and cyber threats. Blockchain technology can eliminate or reduce these concerns by providing the ability to track the availability and movement of assets in real-time, digitally identify assets automatically, verify authenticity of assets at any point in the supply chain, and identify malicious actors.

One significant problem facing the industry today is that it is difficult to track lifecycle information pertaining to a unique part after it leaves the manufacturer. Yet, this information is critical for any part that changes ownership, has a life limit (cycles, hours), has been repaired or modified, or is involved in a rotable exchange program. Blockchain

TABLE 1. Blockchain use case classification.

Tracing & Tracking (49%)		Digital ID (20%)	Certification & Compliance (19%)	Tokenization (7%)	Contracts (4%)
Parts Provenance/ Authentication	Air cargo	Aircraft sales/leases	Continued airworthiness records	Financial settlements	B2B transactions
Temporary import tax	Sensitive cargo	Additive manufacturing/3D printing	Airframe and Powerplant certifications and Repairmen licenses	Airline ticketing	Minimum Equipment List automation
Raw materials sourcing	Baggage handling	Airline maintenance system integration	Pilot qualifications	Cryptocurrency	Contract fulfillment
Counterfeit parts mitigation	Satellite communication/ position updates	Digital Twin/Thread of every aircraft	Safety management	Airline rewards/ loyalty programs	Insurance
Parts tracking	Aircraft communication/ position updates	Digital Thread of parts through lifecycle	Company licenses	Fuel Management/ trading	
Life limited parts tracking	UAS communication/position updates	Embedded digital artifacts	Government licenses	Charitable donations	
Parts exchange/ resell	Air traffic management, airport, and airline coordination	Distributed blackbox	CO_2 emissions offsets		
Rotable /Pooled Parts	Coordination between international boundaries	Human contact tracing	Public records		
Aircraft operational history	Loadable software configuration management	Immunity passports/ travel documents	Accident/incident investigations records		
Supply chain risk management	Hardware configuration management		Maintenance history/ records/MRO		
Inventory optimization	Aircraft deep cleaning		Tooling location and calibration		
Asset management					

technology has the power to create a digital birth certificate for every part that is installed in an aircraft and update it every time the part (or aircraft) is serviced or inspected [14]. The immutability and transparency of the blockchain provides a means to verify ownership, identify the location of the part, and have confidence in the historical data pertaining to its usage and maintenance.

Another problem facing the supply chain is that human intervention is often required to complete transactions. Blockchain technology can eliminate the inefficiencies of having humans-in-the-loop by using automated smart contracts with verifiable digital signatures to issue digital certificates for parts as they move through the distribution channel. The digital signatures would be verified by the stakeholder nodes to prevent spoofing or fraudulent messages and to ensure that all transactions are properly attributed and valid. The result would be reduced turnaround times, optimized inventory levels, reduced occurrences of human error and fraud, and increased resale value of parts whose history are trusted.

A critical issue facing the supply chain is the need to verify the provenance and authenticity of raw materials, new aircraft parts, and rotable aircraft parts. Like many industries, counterfeit parts have made their way into the supply chain. Using blockchain technology to authenticate the source of a part can help to identify unapproved parts that have not undergone the same rigorous certification testing or quality control as approved parts. Not only is this traceability valuable from a safety perspective, it is valuable from a cost perspective in that these unapproved parts are likely to be less reliable. Blockchain technology can also assist in the accurate tracking of recalled or obsolete parts, and the calculation of import/export taxes owed for components that are manufactured in multiple stages in different nations.

AM represents a valuable opportunity to embed digital artifacts into a part so that the information can be used to verify its provenance and authenticity before printing or after production to expedite approvals. The integrity and source of raw materials can be verified through the blockchain, which is particularly valuable for environmentally responsible manufacturing and tracing of materials of concern or conflict minerals. The DoD is also exploring the use of blockchain technology for remote AM on ships at sea and with units in the field [15].

Additive in and of itself is pretty digital. It starts with a model that is transformed digitally and then printed right there. That's definitely prime for enabling the technology. [16]

Krishna Ratakonda, IBM

Cybersecurity concerns have made their way into the supply chain and are of particular concern in the defense industry. According to the Office of the Director of National Intelligence, "US supply chains are under 'constant, systematic assault from foreign entities and other adversaries who look to use the attacks to penetrate sensitive research and development programs, steal intellectual property (IP) and personally identifiable information, insert malware into critical components, and mask foreign ownership, control, and/or influence of key providers of components and services.'" [17] In response, Lockheed Martin became the first US defense contractor to adopt blockchain as part of their approach to cybersecurity by integrating Guardtime Federal Black Lantern appliances and distributed Core Blockchain into their supply chain in 2017 [18].

MRO Challenges

According to an analysis conducted by PricewaterhouseCoopers (PwC), blockchain technology could reduce global MRO costs by 5% or $3.5 billion and increase revenue by 4% or $40 billion [19]. Hence, MRO businesses could significantly benefit if the technology is used to reduce inefficiencies in the logistics chain, optimize inventory levels, and facilitate records management for regulatory compliance.

The potential benefit has not gone unnoticed. For example, in 2017, Block Aero demonstrated the use of the Aviation Blockchain Platform to digitally track the movement and maintenance of aircraft parts through an MRO logistics chain. In February 2020, the industry-wide MRO Blockchain Alliance was announced with similar goals [20]. Both of these peer-to-peer solutions offer participants the ability to track the digital thread of their aircraft part (current status, chain of custody, birth certificate) as well as the digital passport of the part (indisputable identity, ownership, and airworthiness) throughout the MRO logistics chain.

Using blockchain technology to authenticate and manufacture AM parts onsite would eliminate long lead times for ordering parts, costs associated with packaging, shipping, warehousing, inventory levels, and customs. Further, blockchain technology would enable manufacturers to monitor and even limit how many copies of a product or component are printed. This would lessen or eliminate the possibility of, for example, a customer paying to print 50 parts but actually printing 100 and selling the surplus on the gray market. Recently, Moog demonstrated this capability by showing that the lead time of a metal AM part from an F-15 fighter aircraft could be reduced from 265 days to 6 hours, and the lead time of a plastic part could be reduced from 133 days to 1 hour using blockchain technology authentication. The potential value of this capability to the MRO business is undeniably immense.

Blockchain technology has the potential to transform the maintenance and repair side of the industry, not just from the aircraft parts tracking aspect, but from the airworthiness management aspect. The global nature of the industry drives maintenance to be performed all around the world. Contract maintenance is unavoidable and sometimes desirable. Blockchain technology can be used as the integration layer between the airline and the repair organization to correlate maintenance records, facilitate payments, track components and tooling, and ensure personnel are trained and qualified to perform the maintenance. The disparate maintenance systems used throughout the industry make the sharing of maintenance records very difficult. Today, duplication of aircraft records is not unusual: airlines have their own set of maintenance records for each aircraft and repair organizations have their own set of records for the same aircraft, opening the door for potential error in addition to the obvious inefficiency [21]. While not all historical maintenance information should be shared on the blockchain, the information required to perform the maintenance or verify records should be discoverable from the blockchain. Further, information about tooling used in maintenance should be stored in a blockchain so that the location can be tracked, and the calibration status can be verified. Digital identification in combination with verifiable claims could be used to ensure the qualification and training of the personnel performing inspections and maintenance.

Some government mandated inspections and maintenance documents are required by regulators to be preserved for every commercial aircraft, either in service or parked. Records for older aircraft are highly susceptible to being misplaced, incomplete, or having inaccurate maintenance documentation. Currently, many older aircraft are being retired early due to the economic situation related to the COVID-19 pandemic and will be sold or harvested for their parts, creating a secondary market for used aircraft parts. The resale value of these aircraft and used parts will be correlated to their maintenance and usage history. While it is too late to use a blockchain to preserve the historical records of these aircraft and parts, airlines could take advantage of the downturn to build a blockchain platform for their remaining fleet of aircraft, potentially reaping the benefits in the next retirement cycle.

Certification and Compliance Challenges

The management of certification and compliance records is a major issue for the aerospace industry. Blockchain technology could make the authenticity of aircraft records absolute. Maintaining aviation records required for continued airworthiness and compliance audits in an absolutely secure manner was considered an unattainable goal in the age of paper-based records. Blockchain technology can provide the traceability and transparency needed for aircraft records to ensure information is not altered, accidentally or nefariously. In the case of accident or incident investigations, metadata related to the event could be stored in a blockchain until the relevant agency had the opportunity to review it, thus ensuring trust in the immutability of the data.

On June 26, 2020, Pakistani International Airlines (PIA) announced the grounding of one-third of their pilots after determining their licenses were dubious [22]. Subsequently, regulatory agencies around the world revoked PIA's authorization to operate in their national airspace. The investigation into the PIA pilots' licenses was announced after flight PK 8303 crashed in March 2020, killing 98 people. The cockpit voice recorder and transcripts from the control tower seem to indicate pilot incompetence, although the investigation into the crash is still ongoing. Sadly, this event emphasizes the need to ensure that flight crews have the proper credentials and qualifications to operate specific aircraft. The need to ensure the maintenance crews have the proper credentials and qualifications to perform maintenance on the aircraft is also critical to safety. Blockchain technology could be used to ensure training records for all personnel who are in safety-related positions are verifiable and resistant to falsification.

Blockchain technology can be used to ensure the authenticity of electronic Authorized Release Certificates (ARCs). Any new or repaired part intended to be installed on an aircraft must have an ARC attached, such as the Federal Aviation Administration (FAA) Form 8130-3, the European Aviation Safety Agency (EASA)/Transport Canada Form 1, or an equivalent form. Historically, the ARC was a paper certificate stating that the part was airworthy. More recently, Airlines for America (A4A)—formerly the Air Transport Association of America (ATA)—published ATA Spec 2000 Chapter 16, which provided a standard for the electronic version of the ARC in Extensible Markup Language (XML) format. The SAE G-31 Committee for Electronic Transactions for Aerospace is developing a process that would record certain information about the electronic ARC in a blockchain, permitting verification of the full document transmitted via other means. Each ARC record would include a digital signature certifying the airworthiness of the part. As the part is used in service and is subsequently repaired, the maintenance and usage records would be added to the blockchain. Work is already underway at Singapore-based Block Aero to develop a web-enabled blockchain network using Hyperledger Fabric that extracts electronic ARCs from existing systems and records the transaction into their blockchain network. Linking a part's electronic ARCs to its maintenance and usage records would streamline the audit and compliance process as well as enhance the resale value of the part.

Blockchain technology could also be used for environmental compliance programs, such as the Carbon Offsetting and Reduction Scheme for International Aviation (CORSIA). The International Civil Aviation Organization (ICAO) initiative requires airlines from United Nations member states to reduce their CO_2 emissions over a 15-year period starting from a baseline level established in 2019. What this means is that participating airlines operating international flights must monitor, reduce, and report their CO_2 emissions annually. Blockchain technology could provide an automatic and trusted means for participating airlines and nations to track the flight data required to comply with CORSIA.

Cargo Tracking Challenges

According to a 2019 study by Gartner, over $4 trillion in goods are shipped around the world every year [23]. While most of that cargo is shipped via ocean freighters, air cargo plays a major part in the shipment of sensitive merchandise, such as medicines, perishable products, human organs, and live animals. Blockchain technology can transform tracking of air cargo, making it more efficient, providing visibility throughout the process chain, and ensuring that the cargo has not been exposed to damaging environmental conditions. The current air cargo process is very paper intensive, has multiple exchanges between partners, and multiple customs and ports of entry delays. The lack of real-time visibility into location and status of cargo, including the environmental conditions pertaining to sensitive cargo, is a huge issue for the industry. Blockchain technology that uses smart contracts to automate digital signatures can eliminate the inefficient paper processes and provide the needed visibility to expedite transactions, providing tremendous value to all stakeholders. Hence, IATA is developing an initiative called ONE Record, which aims to provide a data-sharing standard for creating a single view of a cargo shipment for any company in the logistics chain [24]. Further, a similar process could be applied to track passenger baggage.

Communications Concerns for Manned and Unmanned Aircraft

Blockchain technology can be used to enhance situational awareness and improve air traffic management of manned and unmanned aircraft around airports and along flight corridors. Improved quality and trust in the operational data, coupled with increased automation, would go a long way in mitigating congestion, ensuring aircraft separation, reducing the risk of incidents, and addressing adversarial contested conditions between international boundaries. In 2017, SITA created the FlightChain project that included multiple airlines and multiple airports. Over two million aircraft flight records were stored and shared between the participants using a private, permissioned blockchain implemented on both the Ethereum and Hyperledger Fabric platforms. The project explored the use of smart contracts to resolve conflicting information [25]. Further research continues regarding how blockchain technology can be used to coordinate the hand-off of aircraft flying in and out of international airspace.

The Automatic Dependent Surveillance—Broadcast (ADS-B) Out mandate, which uses the satellite-based Global Positioning System (GPS) instead of ground-based radar to determine aircraft position [26], went live in 2020. Many industry experts are concerned that the lack of data encryption makes ADS-B vulnerable to cyberattacks, putting aircraft position data at risk for being spoofed or hacked [27]. In anticipation, in 2017 Boeing filed a patent for an anti-spoofing navigation blockchain that uses a system of coordinates to duplicate physical sensors and replace those readings with simulated coordinates in case the physical sensors get hacked [28].

Blockchain technology is the confluence of work that has its roots in decades of cryptography and military strategic communications. As a result, blockchain technology can now be used to protect aircraft and satellite communications from cyberattacks by providing a secure backbone of trusted, unbreakable communication transactions with continuous authentication. The technology can also be used to securely communicate hardware and software configuration and information originating from remote embedded devices onboard an aircraft or satellite.

The increasing demand for larger available bandwidths will impose serious security, safety, and performance concerns in the next years. At the same time, the evolution of unmanned aircraft systems (UAS) and air taxis will use similar communication infrastructures. By 2023, the commercial UAS or "drone" market is expected to exceed 2.9 million aircraft [29]. Some of these vehicles will be operated by commercial remote pilots, while others will be fully autonomous. Either way, these vehicles will be unmanned, meaning that they must be capable of some level of onboard decision-making and off-board communications to ensure the safety of the vehicle and its surroundings. The avionics needed for these vehicles will be highly redundant and will use control laws that are robust (capable of performing without failure under a wide range of conditions) and resilient (capable of recovering from or adjusting easily to internal or external disturbances including cyberattacks). These vehicles will use algorithms and artificial intelligence (AI) for situational awareness and to prevent cyberattacks; however, they will need to communicate with ground systems and other vehicles in order to maintain separation [30].

The anticipated increase in air traffic will raise further concerns over whether successful and secure communication sessions can be established during cyberattacks while maintaining a minimum level of communications services required for survivability. Hence, a blockchain solution that uses a high-resolution time protocol to guarantee accurate timestamps for communications records would be needed [31]. One possible solution already exists: Guardtime Federal blockchain. This blockchain uses a patented distributed calendar infrastructure (as opposed to a DLT infrastructure) to ensure accurate sequencing and timestamps of transactions (records). This blockchain uses a Keyless Signature Infrastructure (KSI) to combine hashes with Merkel trees to create Calendar Trees and Calendar Hash Chains. The blockchain is highly scalable, secure, sequential, and perpetual.

Space Traffic Awareness Concerns

Space traffic awareness and management are becoming significant issues as more nations launch, orbit, maneuver, and deorbit satellites. Blockchain technology can be used to increase coordination and automation of processes to address the need for better space situational awareness and international coordination. Currently, ground sensors are used by private companies, governments, and nations to track the movement of satellites. Data collected from these ground sensors is recorded in each independent space catalog, making it difficult to detect anomalies or discrepancies. To address this concern, the MITRE Corporation has proposed a decentralized, open source blockchain called Blockchain Enabled Space Traffic Awareness (BESTA), which would automatically reconcile these independent space catalogs via a distributed ledger [32]. Once the space catalogs were reconciled, further automation would be possible to enable anomaly detection and to convene appropriate parties to resolve any issues discovered. Adoption of such an international space blockchain, however, would require agreement of stakeholder nations to fund, build, test, and govern the capability. Figure 2 illustrates the types of data transactions that could be shared in the BESTA blockchain.

Digital Identity Concerns for Digital Twins

Digital identity is a fundamental element for establishing a digital thread and a digital twin for a part or an aircraft. The recent data evolution and model-based system engineering trends are moving the aerospace industry toward digital

markets and digital twin solutions. Blockchain technology could be used to securely exchange digital models between collaborating parties during a joint development program or during ownership transfers. The digital models could range from a complex representation (digital twin) of a physical part to an accurate ledger of the history of a part. Moog and ST Aerospace are collaborating on this capability to authenticate the integrity of the digital twin using the Moog VeriPart® solution.

The digital thread of a part could be maintained in the blockchain, starting with a digital birth certificate, and then tracking that part through its lifecycle. For example, the Tracr Association created a digital twin of a natural diamond by storing a 3D scan and digital image of the diamond in their blockchain and then tracking that diamond as it transitioned through the industry value chain [33]. The same technology could be applied to aircraft parts by linking the physical part to the blockchain via digital images and embedded digital artifacts. Research is continuing into crypto anchors for physical parts, such as physically unclonable functions (PUFs), which could be integrated with a blockchain. Tremendous value can be unlocked by using blockchain technology for trusted coordination of processes and assets (physical and digital) between organizations, thus creating a "trusted, cross-organizational digital thread."

Digital Identity Concerns for Tracking People

Blockchain technology can be used to securely issue and verify travel documents and digital passports (including health immunity passports) to enable passenger and crew identification and enhanced immigration processing at

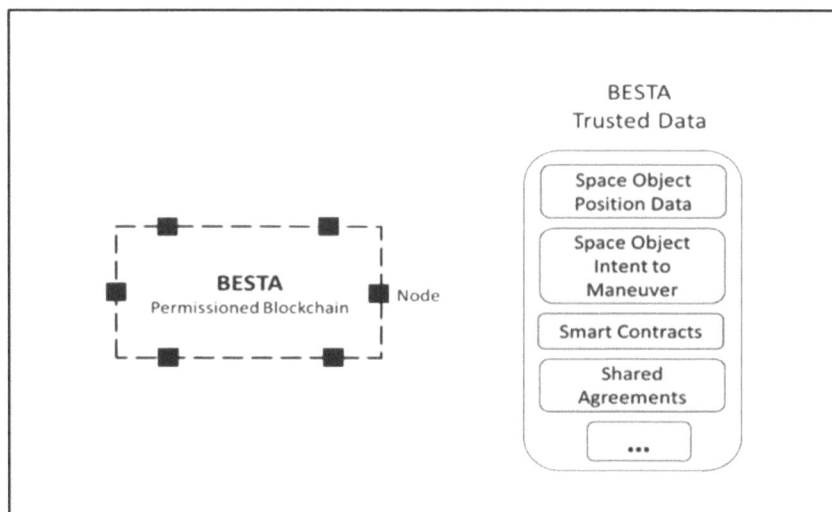

FIGURE 2. BESTA trusted data [32].

ports of entry. For example, SITA Smart Path™ uses digital biometric identity management credentials (facial recognition) to help passengers fast-track through airports, eliminating the need for multiple travel documents or sharing of personal data. Health immunity passports could offer an additional level of confidence at airports, boarding, customs, and immigration.

Prior to the COVID-19 pandemic, tracking of people was primarily focused on tracking work performed by individual persons. The situation changed and society identified a desire to perform contact tracing, which ultimately means that individual people, and everyone they come into contact with, should be tracked. Blockchain technology could be used for tracing exposure to infectious diseases or for international travel. In June 2020, GE Aviation launched its HealthPass blockchain solution to address COVID-19-related safety concerns pertaining to employees, passengers, and aircraft. The blockchain application aims to restore passenger confidence by allowing them to view cleaning history of the aircraft [34]. In the post-COVID-19 world, blockchain technology may provide greater transparency of people, equipment, and procedures for air travelers, governments, and airlines. The technology could be used to help minimize any future outbreaks and allow for implementation of mitigation steps, such as social distancing to minimize the spread of diseases.

Financial Transaction Issues

Blockchain technology could increase the speed and ease of transactions for any financial settlement, such as airline ticketing and checked bags, and the transfer of passengers from one airline to the next due to a flight cancellation. In addition to the single secure token under development by SITA Lab, the use of cryptocurrency to make purchases and conduct business transactions without the need for banks, currency exchanges, or intermediaries is gaining popularity and acceptance around the world and will undoubtedly find its way into aerospace transactions.

Airlines purchase fuel around the world. Blockchain technology could be used for fuel management and trading. FuelPlus launched eTender, which uses the IATA developed XML standard for fuel tenders [35]. Adding a blockchain layer could allow paperless fuel purchases to be made using cryptocurrency, thus reducing the need for currency conversions and third-party vendors. In addition, blockchain technology could speed up the aircraft fueling process by enabling the pilot to request the fuel directly from their Electronic Flight Bag so that the fuel tank arrives and loads the fuel upon aircraft arrival.

Another popular concept is to use blockchain technology to automate the airline rewards and loyalty programs. Airlines could see a substantial benefit by easing the manual burden of accurately tracking frequent flyer rewards programs, travel on code-share partners, and purchases made for tickets, upgraded seats, and gifts.

Global Contractual Challenges

Blockchain technology uses smart contracts that execute automatically without having a legal system or an external enforcement mechanism in place between stakeholders. Self-executing contracts would contain the terms of the agreement written into the code (smart contract), which would then be distributed across the blockchain network. The execution of the contract and all transactions would be controlled by the smart contract code, without the need for an intermediary, manual entry, or human decision-making. The aerospace industry would benefit tremendously from having the ability to automatically execute contracts globally.

The list of opportunities for using smart contracts in business-to-business (B2B) transactions in aerospace is endless and could be as simple as purchasing a part to as complex as making an insurance claim. The downside is that overengineered and complex smart contracts could lead to failures caused by a lack of understanding of this specific technology. Hence, academia and blockchain specialists would need to work with the industry to develop common, useful smart contracts that achieve the intended purpose, while minimizing the confusion and complication of single-use software code. The aerospace industry should also learn from past successes and failures with smart contracts, since one of the benefits (and challenges) is that blockchain has no "rollback" capability.

Recommendations

Blockchain technology can be used to solve numerous challenges and concerns in the aerospace industry. The opportunities are so immense that it is difficult for aerospace organizations to decide where to start. To move forward, industry leaders must agree on which problems need to be solved first. Then, they must consider how the technology will fit into their end-to-end digital transformation strategy focused on customer value creation. Each organization should look at their business processes holistically, evaluate the customer experience across different business units, and consider how blockchain technology could be used to provide the most improvement to their business and to the industry. Once the most promising improvement initiatives are identified, they should select the blockchain application that offers the most value to their business and to the industry as a whole. Once this assessment is made, they can move forward to the next steps, which include overcoming hurdles and deciding on an implementation strategy.

Blockchain technology could be used to solve some of the biggest challenges facing the industry by

1. Identifying counterfeit parts, obsolete parts, and ensuring the integrity of raw materials
2. Reducing turnaround time, optimizing inventory levels, reducing the occurrence of human error and fraud, and increasing the resale value of parts

3. Providing visibility into the location and movement of aircraft parts through the supply chain and MRO logistics processes, and reducing inefficiencies caused by duplication of efforts

4. Providing an immutable record of usage and maintenance for life-limited parts

5. Improving the efficiency and visibility of cargo operations

6. Preventing cyberattacks on aircraft and satellite communications with trusted transactions that are continuously authenticated

7. Improving situational awareness and collaboration for air traffic management and space traffic management

8. Establishing a digital twin for each aircraft part and aircraft tail number to enhance engineering supported sourcing and replacement of aircraft parts in MRO

9. Using the technology to create a trusted cross-organizational digital thread

10. Verifying identity and authenticity of parts and design instructions for remote AM and 3D printing, thus reducing lead times and eliminating costs associated with packaging, shipping, inventory management, and customs

11. Using the technology to track pandemic mitigation efforts by airlines, such as social distancing to minimize the spread of disease

12. Acting as an integration layer between airlines and MROs to synchronize and correlate maintenance records, facilitate payments, track components and tooling, and ensure personnel are trained and qualified to perform the maintenance

13. Securely storing a single source of immutable truth for records that are required for airworthiness, investigations, certification, and compliance

14. Securely storing the training and qualification records of flight crews and maintenance personnel

15. Assisting in incident investigations by ensuring critical data or relevant metadata is secured in an immutable manner

16. Increasing the speed and execution of data retrieval and audits/investigations

17. Simplifying the lease return (aircraft redelivery) process, especially in terms of the record checking

18. Using the technology to comply with environmental regulations

19. Using tokens to increase the speed and ease of financial transactions, such as fuel purchases

20. Increasing the speed and execution of contracts and the eliminating manual entry of data, human error, and human decision-making

21. Ensuring contractual requirements for aircraft parts pooling are met

Overcoming Hurdles Related to Perception and Standardization

One of the challenges facing the aerospace industry is the tremendous inertia resisting the adoption of any new technology because the industry is so highly regulated and competitive. A 2018 PwC blockchain survey of 600 executives across all industries indicated that the greatest hurdles to blockchain adoption were [36]

- 48%—Regulatory uncertainty

- 45%—Lack of trust among users

- 44%—Ability to bring network together (lack of standards)

- 41%—Separate blockchains not working together (lack of interoperability)

- 30%—IP concerns

- 29%—Inability to scale

- 20%—Audit/compliance concerns

When the contributors of this report were asked what they believed were the greatest hurdles impeding the adoption of blockchain technology in aerospace, their responses were similar

- 22%—Lack of knowledge and understanding of the technology

- 20%—Infrastructure required to integrate and scale a blockchain

- 20%—Stakeholder trust and willingness to share data

- 15%—Cost and return on investment

- 12%—Lack of industry standards

- 10%—Regulatory acceptance and cybersecurity

The following sections discuss how these hurdles could be addressed through commitment by industry leaders.

Lack of Knowledge and Understanding of the Technology

Aerospace leaders have routinely looked to consulting firms for information and advice about new technologies, such as blockchain. These firms have used hype and the fear of missing out to slowly drive the industry toward certain technologies, such as agile processes and digital transformation. In 2018, Gartner coined ContinuousNEXT, a strategy to provide corporations with the momentum to adapt to change and to develop new practices, capabilities, and creative ways to succeed [37]. Gartner suggested that corporations should fuse digital technology into their products immediately or they will never catch up. The problem with this philosophy is that companies often wind up embarking on Proof of Concept projects without fully understanding why they are doing it and what the potential value could be from the technology, if implemented.

Blockchain, Cloud, and AI are buzzwords that have received a lot of attention in recent years even though many in the industry do not fully understand what these terms mean for their business. Most people associate blockchain with Bitcoin and, hence, fail to understand how the technology works, what its limitations are, and how it could be used effectively in fields other than finance. Concerns over the legitimacy of cryptocurrency have fueled mistrust and misunderstandings of the fundamentals of blockchain technology even though the introduction of smart contracts and DLT have greatly expanded the way the technology can be applied. Further research in the next few years will help to develop blockchain capabilities so that the technology is mature at a commercial scale and industry leaders can visualize its practical value.

The internet is full of myths about what information can be stored in a blockchain, which is misleading the industry about the real capabilities of a blockchain application. In reality, the more data stored in the blockchain network, the slower the transaction times to achieve consensus across stakeholder nodes, and the higher the cost to run the blockchain due to increasing energy costs. The amount of information stored on the blockchain should be limited to metadata or restricted to the information required to validate a transaction or record. If additional data is needed, digital identification can be used to allow access to information on as-needed basis via links to an external, secure "off-chain" repository. For example, large data sets, such as aircraft sensor data and data used by onboard processors, should not be stored in a blockchain. Instead, this data should be stored securely off-chain with access controlled by the blockchain governance. If this data can be transmitted to the ground separately from the normal aircraft communications channels, data ownership issues could be avoided, and a new revenue stream could be created by the blockchain governor.

For use cases involving life-limited parts (LLPs), the industry is not aligned on what information should be retained for each part. Industry stakeholders want more information (higher resale value), while regulators require only minimal data be kept. For a blockchain to be used successfully for LLP tracking, the stakeholders must agree on what data should be stored, how it should be stored, and how it should be extracted. Ideally, stakeholders will reach agreement on the data exchange before any integration application programming interfaces (APIs) are written between the blockchain and the various Enterprise Resource Planning (ERP) systems at the operators and MROs. The IATA Aircraft Leasing Group released guidance for LLP tracking in their June 2020 report titled "Guidance for Material and Best Practices for Life-Limited Parts (LLPs) Traceability." This document provides a methodology and a template for capturing the information needed to accurately trace an LLP throughout its lifecycle [38]. This template would be the ideal record of an LLP to store on a blockchain.

So why is there confusion? Very little economically sound academic research has been conducted about blockchain technology. Much of the information available on the internet about the technology is written by blockchain solution providers or those who are set to gain financially from a blockchain application. To overcome this hurdle, professional certification programs and continuing education courses are needed to train the workforce about the technology and how to use it to meet specific business needs for aerospace.

Infrastructure Required to Integrate and Scale a Blockchain

Decades of mergers and acquisitions have left the aerospace industry struggling to maintain fragmented ERP systems and product life management (PLM) systems that are not integrated with each other and are not aligned with the latest digital technology. The industry's efforts to migrate to newer technologies have been never-ending and expensive, and hindered by fear of a failure that could jeopardize their authorization or ability to operate. There are roughly 20 different maintenance planning systems used throughout the industry today. Some are quite old, such as Sceptre™, which was created by North Central Airlines 40 years ago, and Airline Maintenance & Operational Systems or "AMOS," which was created by Swiss International Airlines 30 years ago. Today, Electronic Data Interchange (EDI), back-end systems, and homegrown APIs are used to integrate these antiquated maintenance planning systems with other ERP and PLM systems, but the end result often leads to a lack of visibility into the supply chain and MRO logistics chain.

Instead of replacing these fragmented systems with newer technologies or duplicating data in a central data lake, blockchain technology could be used as an integration layer between these existing digital infrastructures. However, a blockchain solution that replicates the current ERP and PLM systems without considering how the end users would interact with the blockchain and use the data would not solve any of today's problems. The blockchain solution must include a utility layer that uses open source software as the interface between existing systems and the blockchain. The design of the blockchain solution must focus on how the data flows, how it is used, who is using it, what purpose the data is serving, and what are the users' transactional needs.

With blockchain, most focus is put on the technology. Yet, in reality, perhaps 10% of the work is related to technology. The other 90% is rethinking the underlying processes and these in the end are all about people. [39]

**Alan Cabello, Allianz Global
Corporate & Specialty**

As aerospace companies strive to become more digital, a well-defined scalable technology strategy is needed to adapt to changing business needs. Including blockchain in the technology framework would require that companies consider what type of blockchain solution is needed for each objective. Companies would likely find themselves in a situation where multiple blockchain solutions and alliances would be required to meet all their business needs. Hence, the digital infrastructure must be able to integrate with different blockchain technologies. This has to happen while still protecting company's private data and IP from external stakeholders and while still ensuring that data is exposed for internal purposes.

Stakeholder Trust and Willingness to Share Data

Two companies conducting business with each other would probably not need a blockchain network since there would likely be a level of trust between them, allowing them to exchange information via shared databases or spreadsheets. As the number of companies requiring access to the same information increases, eventually that number will reach a critical mass where a blockchain application could be the most efficient solution to increase trust, prevent human error, reduce duplication of data, and automate and accelerate processes. However, the situation may arise where Stakeholder A trusts Stakeholder B, Stakeholder B trusts Stakeholder C, but Stakeholder A does not necessarily trust Stakeholder C. Hence, a governance framework, such as a neutral third party, may be needed to manage stakeholder trust.

Over the past decade, data has been considered "the new oil." Today, companies are drilling for this new oil and then putting a firewall around it to monetize it. Because of this strategy, some stakeholders fear that sharing data in a blockchain would reduce their opportunity to create a future revenue stream around that data. Unfortunately, if this stakeholder happens to be the sole source of a critical piece of information and they choose not to share it in the blockchain, the impact could be that an incomplete transaction record is appended to the blockchain. (If this data is related to a specific physical product, the value of the product itself could be reduced or completely eliminated as a result.) In essence, the trustworthiness of the blockchain would be greatly reduced because it could not be relied on to be the complete and accurate source of information. This potential scenario is another reason why a solution that requires metadata to be stored in the blockchain rather than actual data could be more palatable to stakeholders.

A clearly defined, common benefit for using a blockchain application with a governance framework is needed to achieve buy-in from all stakeholders. Creating a governance framework; determining and regulating how the information on the blockchain is stored, accessed, and used to obtain these common benefits; and why the blockchain is the appropriate solution is essential to ensuring stakeholder trust.

Cost and Return on Investment

During economic downturns, return-on-investment (ROI) assessments and cost-benefit analyses are more closely scrutinized as businesses strive to make wise financial decisions. The business model and ROI for a blockchain application is not well-defined and the real cost of implementation is likely dramatically underestimated due to the cost of integration, accessing data in legacy systems, operating across distributed network partners, and ensuring the data is tagged in a consistent manner. The biggest issue across the industry with respect to data identification, storage, and exchange is that there are huge semantic interoperability constraints that were created by data stored internally in silos and by company mergers and acquisitions. These existing issues would need to be mitigated as much as possible before implementing a blockchain solution, and the cost to resolve them would not be cheap.

As with many great technologies, blockchain technology has been hyped as a solution to some problems, which it was never meant to solve. For some use cases, wide-scale adoption by the industry would be required to achieve the full benefit of the blockchain. Early adopters of the technology may not see as much benefit as those who join late, hence, potentially reducing the incentive to be the first to market. The challenge to the industry is to select the right use cases and objectives so that the benefits of the blockchain technology can outgrow the costs of implementing it. Once wide-scale participation and commitment are achieved, the blockchain technology must be matured and verified. The process of maturing and verifying an aerospace system is significant; the blockchain solution must be proven at an international level, each nation would likely operate their own nodes. Central funding would probably not be available, which means each participant would need to fund their internal efforts and agree to a common change management approach. Hence, an industrial consortium where participants pool their resources may be the most viable method to sustain such a blockchain solution. Consortia and alliances are discussed in greater detail in the following section on implementation.

Despite the economic downturn following the COVID-19 pandemic, corporations already invested in digital transformation initiatives will likely continue the work that was previously started. What is likely to change is how corporations approach and invest in new technologies. The aerospace industry will continue to seek new ways to sign and encrypt data, achieve autonomous consensus, use adaptive, prescient AI, and apply predictive analytics to information. Blockchain technology could offer solutions to problems that the industry never faced before in terms of social distancing and aircraft deep cleaning. As companies seek new ways to get passengers back onto airplanes, the industry is likely to see a business case with a documented ROI for contact tracing and health immunity passports.

Lack of Industry Standards

What happens if there are hundreds or thousands of unique blockchain platforms that are not interoperable? As this is the path the industry is currently on, a clear need for standardization has emerged for data exchange between blockchains. A lack of industry-level standards for data and semantic interoperability and blockchain integration is a hurdle the industry is rapidly seeking a solution for. Without these standards, wide-scale adoption of blockchain technology will never happen.

Efforts are underway to leverage existing standards and protocols that were previously developed by A4A, ASM, ICAO, IATA, ISO, RTCA, SAE International, and SITA. However, blockchain technology will require a new set of agnostic standards for processes and workflows. The SAE G-31 Committee is working to define some of these standards while not conflicting with work in progress by IATA, A4A, the Mobility Industry Blockchain Alliance (MOBI), SITA, and so forth.

The contributors to this SAE EDGE Research Report were asked where they perceived standards gaps existed today and their responses are highlighted below.

- **Data interoperability standards** (open source, usage, data exchange, data dictionary, data protocol, data partitioning)

- **Data interface standards** (EDI, XML/JSON files, existing standards, e.g., ATA, Spec 2000, enterprise systems, maintenance systems, APIs, software development kits, B2B data exchange)

- **Governance standards** (data ownership/access/control, multilayer permissioning, flexible/dynamic permission allocation, digital signatures, controlling access to encrypted data, user license agreements)

- **Architecture standards** (identity layer, certificate layer, metadata layer, workflows, processes, autonomous data gathering, rate of transaction throughput, B2B design)

- **Development standards** (requirements, design, implementation, testing, maintaining, training, operating)

- **Libraries of code** (smart contracts, APIs, service-level agreements, distributed applications)

- **Intelligent decision-making standards** (digital twin, machine learning, situational awareness, AI, levels of criticality)

- **Legal/regulatory means to create smart contracts that meet legal and regulatory goals**

All aerospace corporations should be involved in creating these blockchain standards through recognized industry standards groups and trade associations. Further, blockchain adaptation could be regulated through the use of industry standards to guide aerospace engineers on how their end product will be a part of a bigger ledger, and its ledger a part of another one, and so on.

Regulatory Acceptance

Stakeholders have a misconception that data stored in the blockchain will not be accepted by regulators. Engaging regulators, such as EASA, FAA, and Civil Aviation Administration of China, in the standards development process and in the blockchain implementation process will be essential to overcoming this hurdle.

Regulators are more likely to accept maintenance and usage records and digital identification credentials after a reasonable verification and validation process. The more challenging task would be to seek approval to use blockchain technology for air-to-ground communications, on aircraft-critical devices, to counter cyberattacks, to correct for improper pilot input in-flight, or to enforce flight envelope and routing restrictions. These objectives would intersect with safety requirements for airborne equipment, which would drive substantial verification and validation testing and regulatory engagement. To achieve regulatory acceptance sooner, the industry should start with objectives that pertain to record keeping and then advance to solutions that support safety management.

The FAA Enterprise Network Services (FENS) program is in the early stages of planning and acquisition. The FENS program will replace the current telecommunications system that covers 5 million square miles of domestic airspace and 24 million square miles of oceanic airspace [40]. The objectives of the FENS program are listed as follows: maintaining high levels of availability, survivability, security, and performance for critical missions; providing dynamic service provisioning, reconfiguration, and configuration management; providing insight into network service configuration and operations; evolving future technologies; and managing lifecycle costs for communication needs [41]. Blockchain technology could be utilized in many different aspects to achieve these objectives. By engaging the FAA and the various airspace stakeholders early in the design of the FENS, a successful blockchain implementation could be achieved.

Recommendations

Industry leaders must be willing to address hurdles for blockchain technology to be part of a larger solution to global problems impacting the entire aerospace industry. The lack of understanding of blockchain technology—coupled with the need to integrate with existing enterprise systems while transforming for the future "cyber world"—has created an "exploratory stage" hurdle that only a few companies, such as Honeywell and Boeing, have managed to overcome.

To move forward, each organization should take ownership of developing their workforce skills, identify partners that can be trusted to govern and share data, invest in a digital infrastructure strategy that includes blockchain, engage in standards development, and work with regulatory agencies throughout the blockchain development process to achieve regulatory acceptance.

1. Train the workforce with professional certification programs and continuing education courses about blockchain technology and how to use it to meet specific business needs for aerospace.

2. Conduct research guided by hypotheses developed by blockchain alliances to better understand and fully develop blockchain capabilities.

3. Hire cryptographers who are skilled at cybersecurity as well as blockchain technology.

4. Include blockchain in corporate technology roadmaps.

5. Use a governance framework that defines and regulates how information on the blockchain is stored and accessed.

6. Select use cases that are aligned with existing digital transformation strategies to achieve the desired ROI.

7. Engage in creating blockchain standards through recognized industry standards groups and trade associations.

8. Start with objectives that pertain to record keeping and then advance to solutions that support safety management to achieve regulatory acceptance sooner.

Determining the Best Implementation and Governance Strategy

Deciding which blockchain structure to implement is highly dependent on the use case and the stakeholders needed to achieve the desired objectives. The decision is convoluted by uncertainty over what is meant by a centralized blockchain platform versus a decentralized blockchain platform. The concept of centralized could arguably be said to predate blockchain technology and that all blockchains are inherently decentralized since they use distributed ledgers. However, a blockchain structure that has a single entity controlling the blockchain could be described as centralized, and a blockchain structure that does not have a single entity controlling the blockchain could be described as decentralized. The first decision gate when deciding on an implementation strategy is to determine whether the blockchain needs to have a single entity controlling the blockchain and whether or not that structure would assure data integrity and mitigate anticipated attacks or other adversarial behavior. That decision will drive the governance strategy.

Figure 3 illustrates the difference between centralization and decentralization, which implies that all blockchains are decentralized. The blockchain structure will likely influence how many stakeholders and businesses will participate in the blockchain and which use cases are implemented. Figure 4 illustrates the contributors' views when specifically asked about centralized and decentralized blockchains.

FIGURE 3.   **Centralization versus Decentralization.**

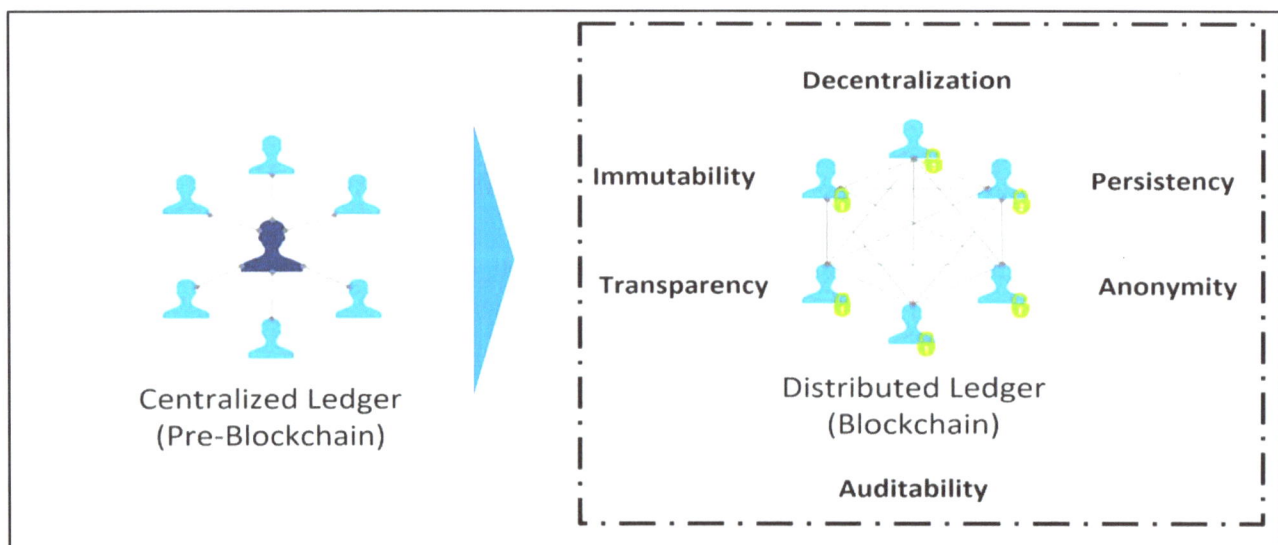

Centralized Ledger (Pre-Blockchain)

Decentralization

Immutability

Persistency

Transparency

Anonymity

Distributed Ledger (Blockchain)

Auditability

FIGURE 4. Blockchain types and use cases.

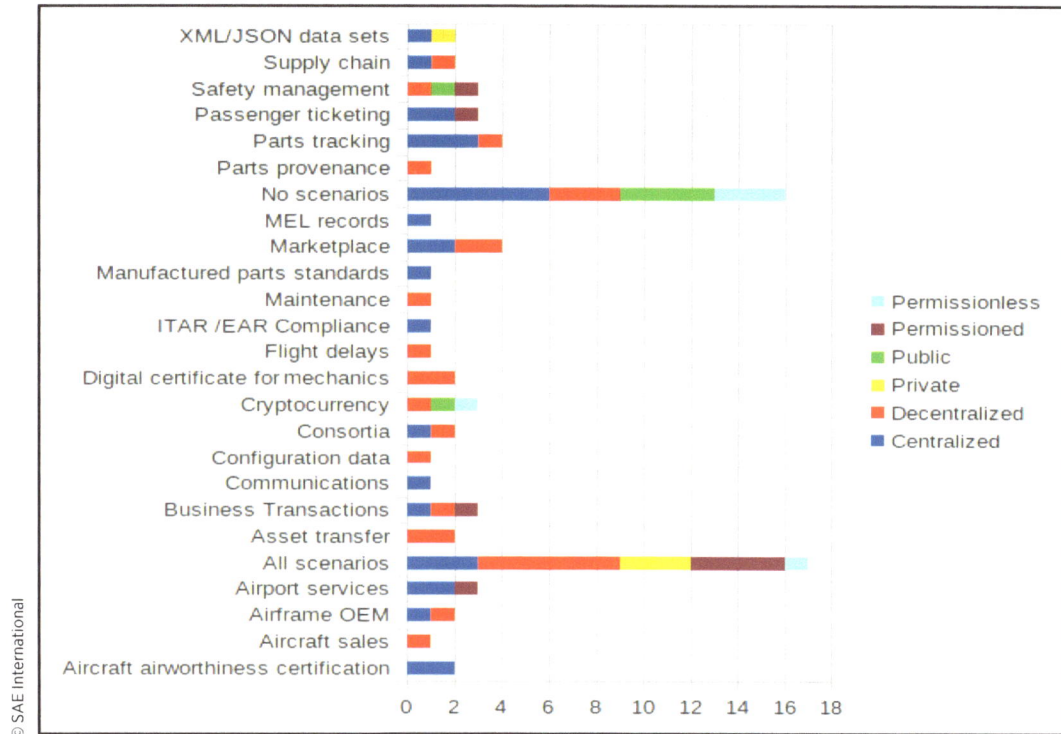

Centralized Blockchain Platform

A centralized blockchain platform can be described as a distributed network of stakeholder nodes that are under the control of a single authority [42]. Hence, a centralized blockchain platform would have a single point of contact, such as Boeing or Airbus. Unlike Figure 3, each stakeholder would have identical copies of the distributed ledger. Proof-of-work or proof-of-stake algorithms would be used, and consensus would need to be reached by all stakeholder nodes before transactions are appended to the blockchain.

Individual companies could have their own centralized blockchains if they needed a single source of immutable truth that could not be achieved via a central database. Alternatively, companies that have disparate enterprise systems that could not be integrated for various reasons, could leverage a centralized blockchain to share data among employees.

The argument against a centralized blockchain is that having all transactions going through a single authority may be too monopolistic and at risk for cyber threats.

Decentralized Blockchain Platform

A decentralized blockchain platform can be described as a distributed network of stakeholder nodes that are not under

the control of a single authority [42]. Hence, a decentralized blockchain platform would not have a single point of contact. No single node or participant would be able to exert authority or influence over the other participants. As shown in Figure 3, each stakeholder would have identical copies of the distributed ledger. Proof-of-work or proof-of-stake algorithms would be used, and consensus would need to be reached by all designated validation nodes before transactions are appended to the blockchain.

A decentralized blockchain platform would be inherently hardened against cyber threats and network outages and would have greater survivability. Should one or more nodes go offline, the remaining nodes and the blockchain would remain fully functional. In fact, the more stakeholder nodes and boundaries involved, the more effective the decentralized blockchain would be in fighting malicious behavior.

Even though the data stored in the blockchain would be tamper evident, the decentralized structure's slow speed and inflexible characteristics could damage the aerospace industry in other ways or could generate new problems that did not exist before the blockchain was implemented. Hence, an incremental approach to deploying a wide-scale decentralized blockchain implementation could help in overcoming these concerns.

Permissioned Blockchain

A permissioned blockchain can be described as a distributed network of stakeholder nodes that has a control layer on top that governs who can participate and what actions they can perform. The limited number of participants (or nodes) ensures that the blockchain is efficient and consensus can be reached quickly. Some permissioned blockchains may have predetermined nodes for validating transactions in order to increase transactional speeds. Permissioned blockchains are also hardened against malicious behavior. For example, permissioned blockchains using byzantine fault-tolerance consensus algorithms could have up to one-third of the nodes offline or malicious and still survive. An example of a permissioned blockchain is Ripple.

Because of the need to ensure safety in aerospace, most use cases would require a permissioned blockchain be implemented with an appropriate governance structure to vet stakeholders and to restrict access to only the information needed to complete their transaction. The blockchain would still have a level of transparency, but not complete transparency so as to protect private data and IP.

Permissionless Blockchain

A permissionless blockchain can be described as a distributed network of stakeholder nodes where any participant can fully participate in the blockchain without being vetted by a governing authority. All participants would be able to view transactions, add nodes, and participate in the autonomous consensus [43]. Bitcoin is an example of the use of a permissionless blockchain.

Even though access to data could be controlled in a permissionless blockchain by using a hash of private data and an external system to maintain control over the private data, aerospace companies involved in the design and manufacturing of aircraft and aircraft parts are unlikely to participate in a permissionless blockchain. However, there may be some financial or niche use cases where a permissionless blockchain could be appropriate when anonymous and transparent data is acceptable to all users.

Private Blockchain

A private blockchain is synonymous with a centralized, permissioned blockchain. An example is Honeywell's GoDirect and iTrace.

Public Blockchain

A public blockchain is synonymous with a decentralized, permissionless blockchain. An example is Bitcoin or any cryptocurrency that uses blockchain technology.

Consortia and Alliances

A consortium or alliance blockchain can be described as a semi-decentralized, permissioned blockchain with multiple stakeholders, usually global, whose participation is governed by a group of approved entities. No single authority has control over the blockchain, which helps to minimize the threat of malicious actors. The number of stakeholder nodes is limited to a specific group, thus ensuring the blockchain is efficient and consensus can be reached quickly. Alliances are well-suited for blockchain technology development because they allow the development costs to be shared across stakeholders who have a common interest and stand to benefit from the technology.

There are numerous alliances already formed, most of which settle on a specific blockchain software platform, such as Hyperledger Fabric or Corda. One of the biggest decisions a corporation poised to adopt blockchain technology will have to make is whether to develop their own private blockchain or to join an existing blockchain alliance that best fits the needs of their business and the objectives they wish to achieve. Ideally, a consortia blockchain would

- Use open source software

- Be agnostic to allow for multiple different types of blockchain technologies and their programming languages to be used

- Act as a single layer technology that could be integrated into each stakeholder's networks

However, to date, this scenario has not been established.

Governance Structure

How should blockchain technology be implemented and who should govern it? The technology implementation and governance strongly depend on the use case selected and the interpretation of centralized or decentralized blockchain. Differing views regarding who should govern the blockchain network also make the decision and conversation between stakeholders more challenging. Who should the industry trust to govern and manage the infrastructure of a blockchain?

Governance directly impacts the success and long-term viability of any blockchain network, making the decision about who should govern it, critical. The governing entity would be responsible for managing stakeholder roles and access, defining minimum requirements that must be adhered to, such as data format and rate of transaction throughput, complying with international rules and export control laws, such as International Traffic in Arms Regulations (ITAR) and Export Administration Regulations (EAR), and securing funding from transactions.

Figure 5 illustrates a blockchain structure for the aerospace and defense industry ecosystem as envisioned by AIA and Boeing [44]. The ecosystem includes government entities, original equipment manufacturers (OEMs), global airlines, small and fixed-based operators, defense organizations, maintenance organizations, suppliers, and many additional stakeholders. At the core of this blockchain structure is central governance.

FIGURE 5. Aviation industry ecosystem [44].

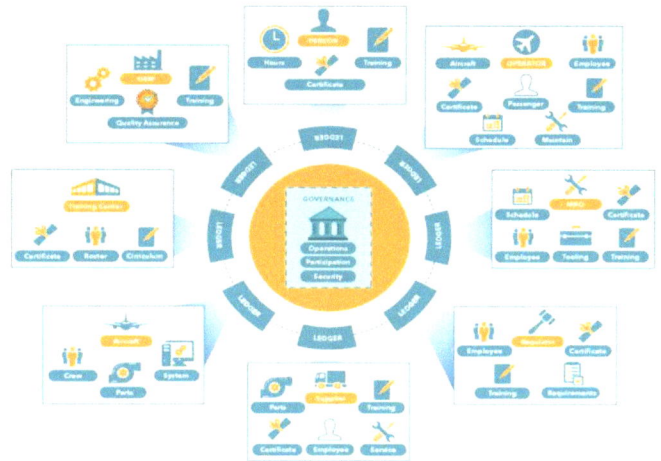

Alternatively, governance could be moderately decentralized, such that multiple stakeholders share the governing responsibilities instead of just one entity. This scenario would enable trust to be established among participants and would create an environment of growth in which additional participants would want to join the network. Alliances often have decentralized governance structures for this reason.

If a single entity were needed to govern the blockchain, the contributors agreed that a neutral third party that represented the entire industry—not just one sector of it—would be ideal. Preferably, the neutral third party would not have an economic interest in the transactions and would view the blockchain as a means to solve a problem, not as a core business.

The FAA and EASA were identified as being appropriate for governing a blockchain for their constituents, although any other regulatory authority would also be appropriate.

While other not-for-profit organizations and trade associations with specific interests should be considered, the following were identified as potential governing entities:

- Airports Council International, "ACI"
- AIA
- A4A
- IATA
- ICAO
- MOBI
- SAE International
- SAE Industry Technologies Consortia®, "SAE ITC®"

While other for-profit organizations with specific business interests should be considered, the following were identified as potential governing entities but were cautioned by concerns over reliance on a single vendor:

- Amazon Web Services, "AWS"
- DXC Technology Corporation

- IBM Corporation
- SAP Corporation
- SITA
- VeriTX

The industry could choose to form its own foundation to act as a neutral governing authority for aerospace blockchain applications. For example, eWINGZ proposed that the industry form an Aviation Network Foundation (AFN) blockchain alliance (Figure 6) that would be funded through low transaction fees to maintain and modernize the network and to conduct research on sustainable global solutions related to aviation.

Recommendations

Blockchain is an important technology having multiple benefits. However, it cuts across organizations, which is uncomfortable for many companies. The aerospace community is confused about who should develop blockchain applications, who should operate them, who should bear the cost, and who owns the data. Once this bridge is crossed, the question of value distribution among the stakeholders/participants will arise. Tremendous value can be unlocked from using the technology.

There will undoubtedly be multiple blockchain networks deployed with very specific objectives to solve very specific problems. Corporations will need to integrate with these networks just to be able to continue to do business. The best way forward is to understand the different blockchain platforms and blockchain types and start by focusing on one implementation strategy that will best solve the most critical problem facing the corporation.

1. Create at least once blockchain alliance, which would allow business risk to be shared across stakeholders, who have a common interest and will benefit from the use cased enabled by the technology.

FIGURE 6. Hypothetical Aviation Network Foundation (AFN).

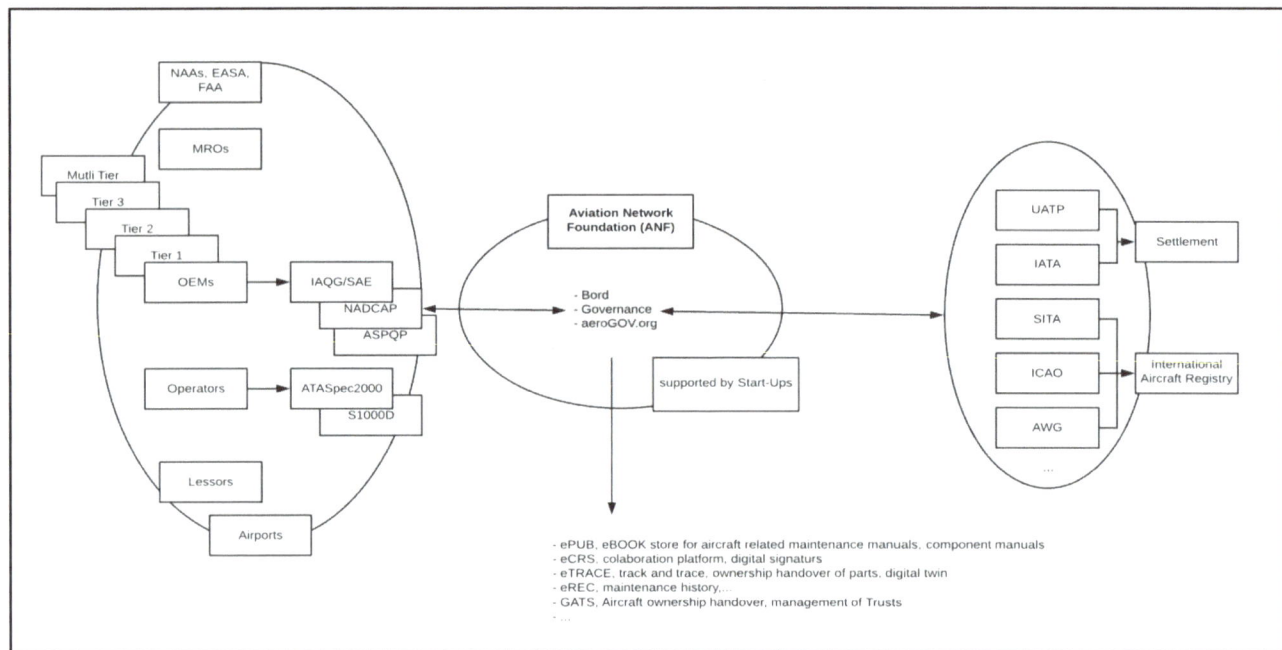

2. Incentivize motivated and innovative stakeholders to join a blockchain alliance by ensuring the governance is decentralized and highly participatory for all stakeholders.

3. Ensure that if a neutral third-party entity is selected to govern (and operate) the blockchain that they are conflict-free.

Summary

Experts believe the use of blockchain technology in the aerospace sector is inevitable and do not want another decade to pass without ample adoption. Some fear that a single event, which could have been prevented by the technology, such as a cyberattack or inadequate pilot training that brings down an airplane (PK 8303 potentially), would have to happen before the industry will fully embrace it.

Blockchain technology is about digital transformation and should be part of a larger technology strategy that has been missing in the industry. This technology provides the opportunity to remove inefficiency of paperwork, have more autonomy, and provide better service for customers. The right blockchain architecture, designed for digital ecosystem integration, could help ameliorate the existing enterprise system fragmentation.

There will not be one blockchain to rule all blockchains. There will likely be as many blockchain networks as there are business networks today. Blockchain technology will not always be the right solution: no technology is the solution to everything. Blockchain technology may be appropriate in some use cases but not in others and may help to streamline critical industry-shared data.

Blockchain technology is immature and may disrupt itself through its own evolution. Just as smart contracts disrupted the blockchain, new value and use cases will drive further changes. The blockchain structure will evolve and could possibly look more like a fishbone rather than a chain and stakeholders may not have a copy of the same data. Further, blockchain technology could be enhanced by quantum computing or it could be completely disrupted by it. A massive cyberattack of quantum computers on a blockchain network could destroy the entire blockchain, depending on the technology used to secure it. However, this risk is no different than any other network risk. Corporations and federal agencies, such as the National Institute of Standards and Technology, will need to invest in blockchain developers who can make the blockchain quantum proof and fortified against these threats.

The unsettled issues that are impeding the adoption of blockchain technology in aerospace are centered around finding the right use cases that solve the greatest challenges, identifying ways to overcome inaccurate perceptions and standardization gaps, and developing the best implementation and governance strategy to achieve the desired outcomes. The technology provides a chance for the industry to come together to share information that is important to everyone as the stakes are very high. The industry needs blockchain, not to compete but rather to come together to move the industry forward, especially in the post-COVID-19 world.

SAE EDGE Research Reports

SAE EDGE Research Reports, like the present report on "Unsettled Topics Concerning Adopting Blockchain Technology in Aerospace," are intended to push further out into still unsettled areas of technology of interest to the mobility industry. SAE launches these reports before attempting to form a joint working group, let alone a cooperative research program or a standards committee.

SAE EDGE Research Reports are intended to be quick, concise overviews of major unsettled areas where vital new technologies are emerging. An unsettled area is characterized more by confusion and controversy than established order. Early practitioners must confront an absence of agreement. Their challenge is often not to seize the high ground but to find common ground. These scouting reports from the frontiers of investigation are intended merely to begin the process of sorting through critical issues, contributing to a better understanding of key problems, and providing helpful suggestions about possible next steps and avenues of investigation.

SAE EDGE Research Reports, therefore, are fundamentally distinct from the more formal working groups approach and far removed from the more mature research program and standard's development process.

Next Steps for Unsettled Topics Concerning Adopting Blockchain Technology in Aerospace

This publication should be considered only as a first step toward clarifying the issues around unsettled topics concerning adopting blockchain technology in aerospace. The intention behind this and other SAE EDGE Research Reports is to start a dialogue among interested parties on important industry-wide topics that require further attention. The expectation is that these explorations of unsettled areas of technology will lead to the formation of working groups and, ultimately, committees that can address and resolve the issues they raise, producing a framework for developing a common vocabulary of definitions, best practices, protocols, and standards needed to support continued progress toward safer and more innovative automotive products.

The experts' collaboration that gave rise to this publication demonstrated a great willingness on the part of the industry to define the terminology, procedures, and eventually the standards needed to enable blockchain technology to move ahead as quickly and efficiently as possible. SAE International has demonstrated its lead in this and closely related areas by the SAE G-31 Committee for Electronic Transactions in Aerospace and its partnership with MOBI.

This SAE EDGE Research Report on unsettled topics concerning adopting blockchain technology in aerospace identifies the following key topics for further pursuit, both through continued informal discussions among industry practitioners and through more formal working groups:

- Use case evaluation and cost-benefit analysis
- Industry standards for interoperability and governance
- Professional training and knowledge exchange
- Implementation strategy

Recommendations

The overall recommendations of this SAE EDGE Research Report could be summarized as follows:

1. Use blockchain technology to identify counterfeit parts, obsolete parts, and ensure the integrity of raw materials.
2. Use blockchain technology to reduce turnaround time, optimize inventory levels, reduce the occurrence of human error and fraud, and increase the resale value of parts.
3. Use blockchain technology to provide visibility into the location and movement of aircraft parts through the supply chain and MRO logistics processes, and reduce inefficiencies caused by duplication of efforts.
4. Use blockchain technology to provide an immutable record of usage and maintenance for LLPs.
5. Use blockchain technology to improve the efficiency and visibility of cargo operations.
6. Use blockchain technology to prevent cyberattacks on aircraft and satellite communications with trusted transactions that are continuously authenticated.
7. Use blockchain technology to improve situational awareness and collaboration for air traffic management and space traffic management.
8. Use blockchain technology to establish a digital twin for each aircraft part and aircraft tail number to enhance engineering supported sourcing and replacement of aircraft parts in MRO.
9. Use blockchain technology to create a trusted cross-organizational digital thread.
10. Use blockchain technology to verify identity and authenticity of parts and design instructions for remote AM and 3D printing, thus reducing lead times and eliminating costs associated with packaging, shipping, inventory management, and customs.
11. Use blockchain technology to track pandemic mitigation efforts by airlines, such as social distancing to minimize the spread of the disease.
12. Use blockchain technology as an integration layer between airlines and MROs to synchronize and correlate maintenance records, facilitate payments,

track components and tooling, and ensure personnel are trained and qualified to perform the maintenance.

13. Use blockchain technology to securely store a single source of immutable truth for records that are required for airworthiness, investigations, certification, and compliance.

14. Use blockchain technology to securely store the training and qualification records of flight crews and maintenance personnel.

15. Use blockchain technology to assist in incident investigations by ensuring critical data or relevant metadata is secured in an immutable manner.

16. Use blockchain technology to increase the speed and execution of data retrieval and audits/investigations.

17. Use blockchain technology to simplify the lease return (aircraft redelivery) process, especially in terms of the record checking.

18. Use blockchain technology to comply with environmental regulations.

19. Use blockchain technology tokens to increase the speed and ease of financial transactions, such as fuel purchases.

20. Use blockchain technology to increase the speed and execution of contracts and eliminate manual entry of data, human error, and human decision-making.

21. Train the workforce with professional certification programs and continuing education courses about blockchain technology and how to use it to meet specific business needs for aerospace.

22. Conduct research guided by hypotheses developed by blockchain alliances to better understand and fully develop blockchain capabilities.

23. Hire cryptographers who are skilled at cybersecurity as well as blockchain technology.

24. Include blockchain in corporate technology roadmaps.

25. Use a governance framework that defines and regulates how information on the blockchain is stored and accessed.

26. Select use cases that are aligned with existing digital transformation strategies to achieve the desired ROI.

27. Engage in creating blockchain standards through recognized industry standards groups and trade associations.

28. Start with objectives that pertain to record keeping and then advance to solutions that support safety management to achieve regulatory acceptance sooner.

29. Create at least once blockchain alliance which would allow business risk to be shared across stakeholders who have a common interest and can benefit from the use cased enabled by the technology.

30. Incentivize motivated and innovative stakeholders to join a blockchain alliance by ensuring the governance is decentralized and highly participatory for all stakeholders.

31. Assure that if a neutral third-party entity is selected to govern (and operate) the blockchain that they are conflict-free.

Definitions

A4A - Airlines for America

ADS-B - Automated Dependent Surveillance—Broadcast

AI - Artificial Intelligence

AIA - Aerospace Industries Association

AM - Additive Manufacturing

API - Application Programming Interface

ARC - Authorized Release Certificate

ATA - Air Transport Association (now A4A)

B2B - Business to Business

BESTA - Blockchain Enabled Space Traffic Awareness

CO$_2$ - Carbon Dioxide

COVID-19 - Coronavirus Disease 2019

DLT - Distributed Ledger Technology

DoD - Department of Defense

EAR - Export Administration Regulations

EASA - European Aviation Safety Agency

EDI - Electronic Data Interchange

ERP - Enterprise Resource Planning

FAA - Federal Aviation Administration

FENS - FAA Enterprise Network Services

GPS - Global Positioning System

IAHM - Integrated Aircraft Health Management

IATA - International Air Transport Association

ICAO - International Civil Aviation Organization

IEEE - Institute of Electrical and Electronics Engineers

IP - Intellectual Property

ISO - International Organization for Standards

ITAR - International Traffic in Arms Regulations

JSON - JavaScript Object Notation

KSI - Keyless Signature Infrastructure

LLP - Life-Limited Part

MRO - Maintenance, Repair, and Overhaul

NIST - National Institute of Standards and Technology

OEM - Original Equipment Manufacturer

PHM - Prognostics and Health Management

PIA - Pakistani International Airlines

PLM - Product Lifecycle Management

PUF - Physically Unclonable Function

PwC - PricewaterhouseCoopers

ROI - Return on Investment

SaaS - Software-as-a-Service

SITA - Société Internationale de Télécommunications Aéronautiques

UAS - Unmanned Aircraft Systems

US - United States

XML - Extensible Markup Language

Acknowledgments

Recognition should go first to all the participants, many of whom also provided feedback on the draft version of this publication. Without their input and initiative, this SAE EDGE™ Research Report would not have been possible.

Tim Abbott, Moog Aircraft Group
Brad balance, Airlines for America
Stylianos Basagiannis, Ph.D., Raytheon Technologies Research Center
Arnaud Brolly, SITA
Dragos Budeanu, International Air Transport Association
Srikanth Challa, Infosys
Paul Conn, Airlines for America
Aharon David, AFUZION-InfoSec
Chris Fabre, Sky Republic
Stefan Fölser, eWINGZ
Leon Gommans, Ph.D., KLM
Fred Jones, Raytheon Technologies Corporation
Jason Jones, Moog Inc.
Ken Jones, Airlines for America
Scott Kordella, Sc.D., MITRE Corporation
Kevin Kuczynski, Pennsylvania State University Applied Research Lab
Ravi Kumar G.V.V., Ph.D., Infosys
Chris Markou, Ph.D., International Air Transport Association
Chuck Marx (Ret.), Digital Aerospace Consulting (PwC)
Sean Melia, SITA
Ravi Rajamani, Ph.D., drR2 consulting
Harvey Reed, MITRE Corporation
Rusty Rentsch, Aerospace Industries Association
Col. James Regenor (jar), VeriTX
Mark Roboff, DXC Technology
Adam Siena, Block Aero
Todd Siena, Block Aero
Aaron Spak, Pennsylvania State University Applied Research Lab
W. Ben Towne, Ph.D., SAE International
Martin Whitfield, SAP Corporation

Additional thanks go to the following individuals who also had the opportunity to provide feedback on the draft version of this publication. Their contributions helped shape this publication.

Satyanaravan Kar, Honeywell
Robert Rencher, Boeing

The author of this document together with the SAE Team responsible for its creation join in expressing our deepest appreciation to all the individuals mentioned above.

Rhonda D. Walthall
Collins Aerospace
SAE International masthead

References

1. Berlin, O., "The Difference between Blockchain & Distributed Ledger Technology," https://tradeix.com/distributed-ledger-technology/, accessed June 20, 2020

2. Applied Research Lab at Pennsylvania State University, 2020.

3. Yuan, S., "Moog and ST Aerospace to Collaborate on Industry's First: Blockchain and 3D Printing-Enabled Total Digital Transaction," ST Engineering, https://www.stengg.com/en/newsroom/news-releases/moog-and-st-aerospace-to-collaborate-on-industry-s-first-blockchain-and-3d-printing-enabled-total-digital-transaction/, accessed Feb. 7, 2018.

4. YouTube, "VeriPartTM Use Case with Moog, Air New Zealand, ST Engineering, & Microsoft," https://www.youtube.com/watch?v=8rD-YEGTE24, accessed Apr. 9, 2019.

5. Kress, A., "Honeywell, iTRACE and SecureMarking Combat Counterfeit Activity in Aerospace with Blockchain," https://www.honeywell.com/en-us/newsroom/pressreleases/2019/12/honeywell-itrace-and-securemarking-combat-counterfeit-activity-in-aerospace-with-blockchain, accessed Dec. 17, 2019.

6. Castillo, M., "Honeywell Is Now Tracking $1 Billion in Boeing Parts on a Blockchain," https://www.forbes.com/sites/michaeldelcastillo/2020/03/07/honeywell-

is-now-tracking-1-billion-in-boeing-parts-on-a-blockchain/#6dd4cc7c78bf, accessed Mar. 7, 2020.

7. Akilo, D., "Airbus Launches Blockchain Platform to Manage Charity and Non-Profit Donations," https://businessblockchainhq.com/business-blockchain-news/airbus-launches-blockchain-platform-to-manage-charity-and-non-profit-donations/, business blockchain, accessed Sept. 29, 2018.

8. MeriTalk, "Blockchain for the DoD Supply Chain?," https://www.meritalk.com/articles/blockchain-for-the-dod-supply-chain/, Dec. 13, 2017.

9. National Center for Manufacturing Sciences, "Project Success: Blockchain Huge Success Story with Five Demonstrations," https://www.ncms.org/project-success-blockchain-huge-success-story-with-five-demonstrations/, accessed Mar. 27, 2019.

10. Ullah, S., "U.S. Department of Defense Looks towards Blockchain for Security," The Tradable, https://thetradable.com/crypto/u-s-department-of-defense-looks-towards-blockchain-for-security, accessed Sept. 2019.

11. Aaronson, M., Caffrey, H., Won, S., and Ahlquist, J., "Getting Real about Blockchain in Aerospace and Defense," https://www.bcg.com/publications/2018/getting-real-about-blockchain-aerospace-defense.aspx, accessed Oct. 2018.

12. Schmidt, J., Gelle, M., and Gottlieb, C., "Driving Trust: Distributed Ledger for Supply Chain," Accenture, https://www.accenture.com/us-en/insights/high-tech/blockchain-aerospace-defense, accessed Jan. 20, 2020.

13. IATA, "Blockchain in Aviation. Exploring the Fundamentals, Use Cases, and Industry Initiatives," accessed Oct. 2018.

14. PwC, "Data for the Life of the Aircraft," https://www.pwc.com/gx/en/industries/aerospace-defence/publications/blockchain-in-aerospace.html, accessed July 14, 2020.

15. MeriTalk, "Blockchain for the DoD Supply Chain?," https://www.meritalk.com/articles/blockchain-for-the-dod-supply-chain/, accessed Dec. 13, 2017.

16. Queen, K. and Brune, B., "In Blockchain We Trust," "SME About," https://www.sme.org/smemedia/sme-media/in-blockchain-we-trust/additive-manufacturing-primed-for-blockchain-adoption/, accessed Jan. 31, 2019.

17. MeriTalk, "Blockchain for the DoD Supply Chain?" https://www.meritalk.com/articles/blockchain-for-the-dod-supply-chain/, accessed Dec. 13, 2017.

18. Lancaster, L., "The World's Biggest Defense Contractor Looks to the Same Tech That Powers Bitcoin for Cyber Security," c/net, https://www.cnet.com/news/lockheed-martin-bets-on-blockchain-for-cybersecurity/, accessed May 2, 2017.

19. Marx, C., Sealy, R., and Thompson, S., "How Blockchain Can Improve the Aviation Industry," Tech & Innovation S+B Newsletters, June 7, 2019.

20. "MRO Blockchain Alliance Launched by SITA and Key Industry Partners," Fifty Sky Shades, https://50skyshades.com/news/maintenance-trainings/mro-blockchain-alliance-launched-by-sita-and-key-industry-partners, accessed Feb. 4, 2020.

21. Dijke, A., "Blockchain—Overview & Applications in the Aviation Industry," Survey Paper for Etihad Airways Engineering, accessed July 24, 2018.

22. Hassan, S. and Shahzad, A., "Pakistan Flag Carrier to Ground a Third of Pilots over 'Dubious' Licenses," Reuters, https://www.reuters.com/article/us-pakistan-airlines-pilots-idUSKBN23W2G3, accessed June 25, 2020.

23. Panetta, C., "The CIOs Guide to Blockchain," https://www.gartner.com/smarterwithgartner/the-cios-guide-to-blockchain/, accessed Aug. 1, 2020.

24. IATA, "ONE Record," https://www.iata.org/en/programs/cargo/e/one-record/, accessed June 5, 2020.

25. SITA, "Flightchain. Research into the Usability and Practicalities of Blockchain Technology for the Air Transport Industry," White Paper, 2017.

26. Thurber, M., "ADS-B Is Insecure and Easily Spoofed, Say Hackers," AIN Online, https://www.ainonline.com/aviation-news/aviation-international-news/2012-09-03/ads-b-insecure-and-easily-spoofed-say-hackers, accessed Sept. 3, 2012.

27. Say, N., "NASA and Aerospace Sector Jump on Blockchain Technology," Blockonomi, https://blockonomi.com/nasa-aerospace-sector-blockchain/, accessed Jan. 15, 2019.

28. Bitnews Today, "Boeing to Patent Anti-Spoofing Navigation Blockchain," https://bitnewstoday.com/news/blockchain/boeing-to-patent-anti-spoofing-navigation-blockchain/, accessed Dec. 18, 2017.

29. Frost & Sullivan, "Commercial Drone Market to Hit 2.9M Units by 2023, Says Frost & Sullivan," https://www.uasvision.com/2020/04/14/commercial-drone-market-to-hit-2-9m-units-by-2023-says-frost-sullivan/, accessed Aug. 1, 2020.

30. Walthall, R., "Unsettled Topics on the Use of IVHM in the Active Control Loop," SAE International Edge ™ Research Report, accessed July 2020.

31. NIST, "Enhanced Distributed Ledger Technology," https://csrc.nist.gov/Projects/enhanced-distributed-ledger-technology, accessed June 20, 2020.

32. Reed, H., Dailey, N., Carden, R., and Bryson, D., "Blockchain Enabled Space Traffic Awareness (BESTA): Automated Discovery of Anomalous Behavior," in *70th International Astronautical Congress*, Washington, DC, Oct. 21-25, 2019.

33. The TRACR Community, "tracr Is Connecting the Diamond Industry by Establishing Provenance, Authencity and Traceability throughout the Entire Value Chain," https://www.tracr.com/, accessed June 26, 2020.

34. GE Aviation, "GE Aviation launches Health Application ID for the Aviation Industry," https://www.geaviation.com/press-release/digital-solutions/ge-aviation-launches-health-application-id-aviation-industry, accessed June 11, 2020.

35. Charalambous, M., "FuelPlus Launches Paperless Procurement Solution for Aviation Fuel," businesswire, https://www.businesswire.com/news/home/20180702005086/en/FuelPlus-Launches-Paperless-Procurement-Solution-Aviation-Fuel, accessed June 6, 2020.

36. pwc global, "PwC's Global Blockchain Survey," https://www.pwc.com/gx/en/issues/blockchain/blockchain-in-business.html, accessed June 20, 2020.

37. Gartner newsroom, "Gartner Says ContinuousNEXT is the formula for Success through Digital Transformation and Beyond," https://www.gartner.com/en/newsroom/press-releases/2018-10-15-gartner-says-continuousnext-is-the-formula-for-success-through-digital-transformation-and-beyond, accessed Oct. 15, 2018.

38. IATA Aircraft Leasing Group, "Guidance Material and Best Practices for Life-Limited Parts (LLPs) Traceability," June 2020.

39. Groenfeldt, T., "Allianz Prototypes Blockchain for Global Self-Insurance Client," *Forbes*, https://www.forbes.com/sites/tomgroenfeldt/2017/11/14/allianz-prototypes-blockchain-for-global-self-insurance-client/#356e1a2578e6, accessed Nov. 14, 2017.

40. Thompson, L., "Raytheon Technologies Leverages Merger Synergies to Mount Bid for FAA Network Transformation," https://www.forbes.com/sites/lorenthompson/2020/07/09/raytheon-technologies-leverages-merger-synergies-to-mount-bid-for-faa-network-transformation/#7cf40d207d59, accessed July 9, 2020.

41. FAA.GOV, "FAA Enterprise Network Services Program," https://www.faa.gov/air_traffic/technology/cinp/fens/, accessed July 9, 2020.

42. Seth, P., "Difference between Centralized and Decentralized Blockchain," https://blogs.systweak.com/difference-between-centralized-and-decentralized-blockchain/, systweak blog, accessed July 9, 2018.

43. SITA, "SITA Flightchain: Research into the Usability and Practicalities of Blockchain Technology for the Air Transport Industry," 2017.

44. Airline Industries Association, "Blockchain in Aerospace & Defense, Establishing an Industry Approach to Blockchain Governance, Standards, and Participation," AIA Whitepaper, May 2019.

Contact Information

EDGEresearch@sae.org.